BULLETIN NO. 11

OF THE

ILLINOIS STATE MUSEUM

OF

NATURAL HISTORY.

NEW SPECIES OF PALÆOZOIC INVERTEBRATES
FROM ILLINOIS AND OTHER STATES.

BY S. A. MILLER AND WM. F. E. GURLEY.

Published quarterly by the Illinois State Museum of Natural History.

SPRINGFIELD, ILLINOIS.

AUGUST 20, 1896.

Entered as second class matter at the Postoffice at Springfield, Ill.

SPRINGFIELD, ILL.
ED. F. HARTMAN, STATE PRINTER.
1896.

ILLINOIS STATE MUSEUM

OF

NATURAL HISTORY

SPRINGFIELD, ILLINOIS.

NEW SPECIES OF PALÆOZOIC INVERTEBRATES FROM ILLINOIS AND OTHER STATES.

BY S. A. MILLER AND WM. F. E. GURLEY.

SUBKINGDOM MOLLUSCA.

CLASS LAMELLIBRANCHIATA.

ORDER SIPHONIDA.

FAMILY CARDIIDÆ.

LUNULICARDIUM GRANDE n. sp.

Plate I, Fig. 1, right valve; Fig. 2, left valve of another specimen; Fig. 3, anterior view of same.

Shell very large, subovate. Valves highly convex or tumid. Height about one-sixth more than the length. Anterior side rounded and recurved toward the basal margin. Posterior and basal margins broadly rounded. Beaks near the middle, prominent and incurved over the cardinal line. Umbones gibbous and gradually merged into the convexity of the valves. Thickness through the greatest convexity of the valves about one-sixth less than the greatest length. Surface marked by from seventy to ninety fine radiating plications, that increase in size without division or implantation, toward the margin of the shell. Our specimens are casts but the plications are so distinct, that without careful inspection, one might suppose he was looking at the shell itself. There are some undefined concentric undulations of growth of the shells preserved on the casts.

This species is so different from any other defined *Lunulicardium* that no comparison with any of them is necessary. It may not be a *Lunulicardium* but as the hinge line and interior are unknown in any species of *Lunulicardium* it is referred to the genus, but it approaches nearer to it in external appearance than to any other genus.

Found by R. A. Blair in rocks of the age of the Hamilton Group at Providence, Missouri, and now in the collection of S. A. Miller.

LUNULICARDIUM RETRORSUM, n. sp.

Plate II, Fig. 26, right valve of large specimen; Fig. 27, left valve of small specimen; Fig. 28, cardinal view of same.

Shell small, obliquely subovate. Valve highly convex. Height about one-fourth more than the length. Anterior side recurved and merged into the basal margin. Posterior margin somewhat truncated and then gently rounded to the post basal extremity. Basal margin abruptly rounded at the post basal margin. Beaks near the middle, very prominent and incurved over the hinge line. Umbones gibbous and gradually merged into the general convexity of the valves. Thickness or width through the greatest convexity of the valves equal to the length. Surface marked by sixty or seventy fine radiating plications that increase by implantation in the umbonal region. There are also undefined concentric undulations of growth of the shells preserved on the casts. Our specimens are casts quite well preserved.

This species is distinguished by the great convexity of the valves, prominent incurving beaks and retrorse or recurved anterior margin.

Found by R. A. Blair in the Chouteau limestones, near Sedalia, Missouri, and now in the collection of S. A. Miller.

BLAIRELLA, n. gen.

Shell equivalve, inequilateral, elliptical, subovate or subcircular. Margins closed. Beaks anterior to the middle and incurved. Umbones high and merging into the general convexity of the shell. below. Cardinal line straight posterior to the beaks. Ligament external and contained in a groove along the cardinal line. No cincture on the sides of the valves. Surface marked by concentric undulations or concentric lines of growth. There is a concave pit beneath the beak of the right valve, anterior to which there is a single tooth and posterior to which there is a strong bifid tooth. Muscular impressions and pallial line not observed. Type B. sedaliensis.

In general form this genus most resembles *Edmondia* and if we had not found the hinge teeth and hinge line we would have contented ourselves by referring the species to that genus. But *Edmondia* has a narrow hinge and no teeth; while this genus has a wide hinge, three teeth and a pit beneath the beak, which separate the genera into distinct families. *Edmondia* is classed with the *Cardiomorphidæ* while this genus, by its internal structure, allies itself with radiately ribbed shells and may provisionally be referred to the *Cardiidæ*. The generic name is in honor of the veteran naturalist and collector R. A. Blair, of Sedalia, Missouri.

BLAIRELLA SEDALIENSIS, n. sp.

Plate I, Fig. 4, hinge line; Fig. 5, outer side of the same shell, which is eroded; Fig. 6, right valve of a cast; Fig. 7, right valve of another cast; Fig. 8, Cardi nal view of the same specimen.

Shell subovate to subcircular. Length from one fifth to one fourth greater than the height. Cardinal line straight, posterior to the beaks, and more than half the length of the shell. Anterior, posterior and basal margins rounded. The posterior margin a little more acutely rounded than the anterior, and the basal margin more broadly rounded than either of them. Valves regularly convex and somewhat gibbous in the umbonal region. Width or thickness through the valves nearly equal to half the greatest length of the shell. Beaks forward of the line of the anterior third, prominent, and incurved over the hinge. Post cardinal and antero-cardinal slopes concave. Shell thick on the cardinal line and at the anterior end. Cardinal line grooved externally for the ligament. Large concave pit beneath the beak of the right valve, anterior to which, there is a long tooth, and posterior, to which, there is a strong bifid tooth. Surface marked by very numerous concentric lines of growth.

Found by R. A. Blair and also by S. A. Miller, in the Chouteau limestone, near Sedalia, Missouri, and now in the collection of S. A. Miller.

Family CARDIOMORPHIDÆ.

EDMONDIA ALBERSI, n. sp.

Plate I, Fig. 9, right valve; Fig. 10, left valve of another specimen; Fig. 11, cardinal view of same specimen with point of beak of left valve broken off, and beak at right valve broken off.

Species rather large, six specimens at hand, one smaller and one larger than either one illustrated. Shell subcircular, our specimens are casts, and the length and height are subequal. Cardinal line is gently curving, anterior end somewhat sharply rounded, posterior end subtruncate, shell most produced at the posterior lower margin. Beaks a little forward of the central part, high, and curving over the cardinal line. Umbones high, anterior and posterior sides subangular, the angularity merging below into the general convexity of the shell. Valves somewhat gibbous in the middle part. Pallial line curves quite regularly and is near the margin. Surfaces of the casts retain the evidences of five concentric lines on the shell.

We have little doubt about the generic reference of this species, though the hinge is unknown, and it bears no near resemblance, in form, to any other described species.

Found by R. A. Blair, in the Chouteau limestone, at Sedalia, Missouri, and now in the collection of S. A. Miller. The specific name is in honor of Mr. A. Albers, a very good palæontologist as well as an excellent artist.

CHÆNOMYA LONGA, n. sp.

Plate I, Fig. 12, cardinal view; Fig. 13, left valve.

Species medium size. Shell more than three times as large as high. Cardinal line concave, less than half the length of the shell. Ligament external. Anterior end narrowly rounded and closed. Posterior end widely gaping, dilated and produced nearly half the length of the shell, from the end of the hinge line, to the postero-basal margin. Basal margin nearly straight, though slightly constricted in the middle part. Beaks near the anterior end, somewhat acute and incurved. Umbonal area depressed and from which arises a broad undefined cincture, that extends to the basal margin. Post umbonal slope angular, but gradually loses the angularity, posterior to the end of the hinge line, and merges into the general convexity of the shell, toward the post-basal extremity. Dorsal side of each valve flattened and inclined from the post umbonal slope to the external ligament. Surface marked by strong concentric lines of growth below the post umbonal slope but between that and the hinge line the concentric lines are much smaller.

This is a marked species and no comparison with any other is necessary.

Found by R. A. Blair, in the Chouteau limestone, near Sedalia, Missouri, and now in the collection of S. A. Miller.

SPHENOTUS SINUATUS, n. sp.

Plate II, Fig. 29, right valve of large specimen; Fig. 30, left valve of a smaller specimen believed to belong to the same species.

Specimens variable, in size, from one-half as large as the smaller one illustrated to one-fourth larger than the larger illustration. Shell trapezoidal. Cardinal line straight or slightly arched. Anterior end gently rounded below the beaks. Posterior margin obliquely truncate and then rounded into the basal margin. Basal margin broadly and slightly constricted in the middle part. Valves depressed convex on the sides but more gibbous in the umbonal region. Length twice as great as the height. Beaks at the anterior end, small, flattened and appressed. Umbones angular from the beak down the posterior slope to the post inferior extremity. Post cardinal slope convex and marked along the center by a median angular ridge, in some specimens, and by two median ridges separated by a concave furrow in others, which extend to the lower end of the oblique truncation of the posterior margin. On our larger specimens, which we regard as the types of the species, there are two median ridges, on the posterior cardinal slope, and only one, on the smaller specimens. Cincture oblique and extends from the beak to the constriction in the basal margin. Surface marked by concentric lines of growth. All of our specimens are casts. No part of the shell is preserved.

This is a marked species and no comparison with any other is necessary.

Found by R. A. Blair, in the Chouteau limestone, near Sedalia, Missouri, and now in the collection of S. A. Miller.

—2

ORDER ASIPHONIDA.

FAMILY ARCIDÆ.

MACRODON FACETUS, n sp.

*Plate I, Fig. 14, cardinal view; Fig. 15, right valve, both ends
of the specimen are broken; Fig. 16, surface of the
cast magnified six diameters.*

Species small Our specimens are casts but part of the surface
markings of the shell are preserved. Shell elongate, about or
more than twice as long as high; highest about the anterior third.
Valves convex and thickness about equal to the height. Cardinal
line straight and a little short of the greatest length of the shell.
Anterior end angular at the cardinal line and then rounded into
the basal margin. Posterior end obliquely truncated, from the
end of the cardinal line, and then narrowly rounded, at the post-
basal extremity, into the basal margin. Basal margin broadly and
somewhat evenly rounded. Beaks near the anterior end and incurved above the hinge line. Umbones prominent and fading
into the general convexity of the shell. Post-umbonal slope subangular and extending to the post-basal extremity. No cincture.
Cardinal slope concave. Surface marked by concentric lines of
growth that are plainly visible to the unaided eye, and by fine
radiating striæ that give a beautiful cancellated structure to the
cast under an ordinary magnifier. This surface ornamentation is
doubtless plainly visible on the shell itself.

This species will be readily distinguished by its general form,
obliquely truncated, posterior end, concave cardinal slope and
surface ornamentation.

Found by R. A. Blair in the Chouteau limestone, near Sedalia,
Missouri, and now in the collection of S. A. Miller.

MACRODON PETTISENSIS, n. sp.

*Plate II, Fig. 17, right valve; Fig. 18, right valve of a larger
specimen, with the anterior end broken off.*

Species below medium size. Shell subelliptical or subovate,
wider behind. Length about one and two-thirds the height.
Cardinal line straight and almost equaling the greatest length of
the shell. Anterior end gently rounded into the basal margin.
Posterior end somewhat subtruncated in the upper part and

rounded into the basal margin. The basal margin in the subelliptical specimens is broadly rounded, but, in the subovate specimens, the postero-basal margin is somewhat produced and more abruptly rounded. Valves compressed or only moderately convex. Beaks a little anterior to the middle, obtuse and rising slightly above the cardinal line. Umbones gently convex with a slightly angular posterior slope that merges into the general convexity of the shell near the postero-basal margin. Shell marked by regular concentric lines of growth, and a few obscure radiating lines on the posterior umbonal slope.

This species most resembles *Macrodon hamiltoniæ*, but is distinguished from that species by being shorter in proportion to its length, subangular posterior umbonal slope, and the absence of radiating lines, except obscure ones on the posterior umbonal slope. The surface ornamentation is a prominent feature in *M. hamiltoniæ* while the lines on this species are quite inconspicuous, though our shells are finely preserved.

Found by R. A. Blair, in the Chouteau limestone, near Sedalia, Missouri, and now in the collection of S. A. Miller.

MACRODON BLAIRI, n. sp.

Plate II, Fig. 19, right valve of a large specimen; Fig. 20, left valve; Fig. 21, left valve of a small specimen; Fig. 22, cardinal view; Fig. 23, left valve of same showing some of the shell at the posterior end; the basal margin is eroded; Fig. 24, left valve of a medium specimen.

Specimens in this species variable in size, from small to large. Shell elongated, about twice as long as high. Highest near the anterior third. Valves convex, thickness about two-thirds the height. Cardinal line straight and constituting the greatest length of the shell. Anterior end angular at the cardinal line and gently recurved to the basal margin like the prow of a boat. Posterior end truncated to near the post-basal margin and then abruptly rounded into the base; sometimes the cardinal line terminates in an acute point and the truncated end is incurved to near the post-basal margin. Basal margin broadly rounded or nearly parallel with the cardinal line in the middle part. Beaks near the anterior third of the shell, somewhat acute and rising above the hinge line. Umbones slightly depressed, with an undefined sulcus near the beaks, which fades out on the convex part of the shell, and does not produce a constriction at the basal margin. Post-umbonal slope subangular and extends to the post-basal margin. Surface marked by concentric, lamellose lines of growth and finer radiating striæ that are plainly visible on the casts, to the unaided eye, but on the shell itself presents a beautiful cancellated ornamentation.

This is a very handsome and marked species that cannot be mistaken for any other.

Found by R. A. Blair and S. A. Miller, in the Chouteau limestone, near Sedalia, Mo., and now in the collection of S. A. Miller. The specific name is in honor of the veteran collector, R. A. Blair.

Family AVICULIDÆ.

POSIDONOMYA LASALLENSIS, n. sp.

Plate 1, Fig. 17, left valve; Fig. 18, same magnified two diameters.

Species medium size. Shell subovate. Anterior margin obliquely truncated in front of the beaks, and then vertically, so as to leave a small ear in front, and then broadly rounded, which is continued regularly into the basal margin. Hinge line straight posterior to the beaks, slightly elevated and terminates in an obtuse extremity. Posterior margin below the wing broadly rounded which is continued regularly into the evenly rounded base. The posterior wing is flat, and separated from the body of the shell by an oblique undefined sulcus. Beak anterior to the middle of the shell, acute but not projecting much if any above the hinge margin. Umbones convex, and merging into the general convexity of the shell. Surface marked by six or seven distant, elevated, concentric rounded ridges that do not appear as concentric undulations of growth, but as distinct lines of surface ornamentation. Between these rounded ridges there are numerous fine concentric lines.

This species is so different from all other forms that have been referred to *Posidonomya* that it is with some hesitation we refer it to that genus. It is an aviculoid shell and seems to be nearer to that genus than to any other in the family *Aviculidæ*.

Found in the Coal Measures at La Salle, Illinois, and now in the collection of Wm. F. E. Gurley.

LIOPTERIA SUBOVATA, n. sp.

Plate II, Fig. 7. left valve; part of the ear broken away; Fig. 8,
right valve of another specimen, Fig. 9, left valve
of another specimen.

Species medium size, subrhomboidal. Body obliquely subovate. Anterior margin nearly straight above and abruptly rounded into the basal margin below. Basal margin narrowly rounded. Posterior margin somewhat straight above and abruptly rounded into the basal margin below. Hinge line straight from the anterior side of the beak to the posterior extremity, and nearly as long as the shell. Beak of each valve obtuse and situated near the anterior side of the shell. It is separated from a short ear by an undefined longitudinal sulcus. Both valves are moderately gibbous, the left valve rather more convex than the right. The posterior umbonal slope is rounded and soon merges into the general convexity of the shell. The wing is flat and terminates in a long acute extremity. The shell is marked with concentric lines of growth.

All of our specimens are casts and the left anterior ear is always injured. No radiating lines are discernable.

This is a shorter form with a larger wing than *L. speciosa*, and its general outline will distinguish it from all other species.

Found by R. A. Blair, in the Chouteau limestone, near Sedalia, Missouri, and now in the collection of S. A. Miller.

LIOPTERIA SPECIOSA, n. sp.

Plate II, Fig. 10, left valve, slightly broken anterior to the beak.

Species rather above medium size, subrhomboidal. Body narrow above and obliquely elongate ovate. Anterior margin broadly rounded above and abruptly curving into the basal margin below. Posterior margin nearly straight until it rounds into the basal margin. Hinge line straight from the anterior side of the beak to the posterior extremity and about half the length of the shell. Beak of the left valve obtuse and situated near the anterior end of the shell. It is separated from a short ear by an undefined longitudinal sulcus. The umbonal region is somewhat gibbous, but the greater gibbosity of the valve is in the middle part. The posterior umbonal slope is subangular above but gradually fades away below and is lost in the general convexity of the shell. The wing is small, flat and has an acute extremity. Shell marked with distant concentric lines of growth.

All of our specimens are casts of the left valve, no part of the shell is preserved, and no radiating lines are discernable.

This species is distinguished by its oblique, elongate-ovate form and small posterior wing.

Found by Miss Jessie Blair, in the Chouteau limestone, near Sedalia, Missouri, and now in the collection of S. A. Miller.

FAMILY MYTILIDÆ.

MYTILARCA JESSIEÆ, n. sp.

Plate II, Fig. 1, left valve nearly complete; Fig. 2, another left valve with more of the anterior end broken off; Fig. 3, part of another left valve; Fig. 4, anterior end of right valve complete; Fig. 5, anterior end of left valve complete, small specimen; Fig. 6, part of the surface of the shell magnified four diameters.

Our specimens are quite variable, from medium to considerably above medium in size.

Shell oblique, elongate, subelliptical, with venticose valves in middle part, where it is subcylindrical. The diameter through the valves in the middle area, is nearly equal to the distance from the hinge to the basal margin. Posterior end of the shell cuneiform and narrowly rounded. Basal margin of the shell oblique, nearly straight or slightly arcuate. Hinge line about half the length of the shell, slightly arching. Posterior to the end of the hinge, the shell is slightly arcuate until it abruptly curves into the post-basal margin. Beaks obtuse and terminal at the anterior end. Umbones high and abrup'ly rounded to the hinge line, posterior to which the umbonal slope merges into the cuneate form of the shell. Surface marked by concentric lines of growth. Shell thin.

This species is readily distinguished by its general elongate form from all others. We have numerous fragments of this species but most of them are casts. Part only of the shell is preserved on a few of the fragments.

Found by R. A. Blair, S. A. Miller and Miss Jessie Blair, in whose honor we have proposed the specific name, in the Chouteau limestone, near Sedalia, Missouri, and now in the collection of S. A. Miller.

Family MODIOMORPHIDÆ.

ELYMELLA MISSOURIENSIS, n. sp.

Plate II, Fig. 11, left valve; Sig 12, cardinal view of a cast.

Species medium size. Shell subelliptical, narrower in front. Length twice as great as the height. Cardinal line straight posterior to the beaks, but abruptly dropping in front. Anterior end narrowly rounded. Posterior end broadly rounded. Basal margin rounded more gently at the anterior than at the posterior end. Lunnule deeply impressed. Posterior cardinal slope concave. Valves convex and somewhat gibbous in the umbonal region. Beaks at the anterior end rounded and incurved over the cardinal line. Umbones high and broadly rounded. Posterior umbonal slope subangular at first, but the angularity gradually fades away and the umbonal ridge is merged into the general convexity of the shell. Surface marked by fine concentric striæ growth.

There may be some doubt about the reference of this species to *Elymella*, as it is quite distinct from the forms Hall referred to that genus. But we think it resembles *Elymella* more than *Modiomorpha* and it certainly belongs to one of the genera. In fact Hall only regarded *Elymella* as a subgenus of *Modiomorpha* but we prefer to treat them at present as distinct genera.

Found by R. A. Blair in the Chouteau limestone, near Sedalia, Missouri, and now in the collection of S. A. Miller.

CYPRICARDELLA EXIMIA, n. sp.

Plate II, Fig. 25, left valve with posterior end broken off.

Species medium size. Shell subrhomboidal to subtriangular. Length one-half greater than the height. Anterior end abruptly truncated anterior to the beaks and then prolonged at the antero-basal extremity and abruptly rounded into the basal margin. Cardinal line arcuate. Posterior margin obliquely truncated, prolonged at the post basal extremity and rapidly rounded into the basal margin. Basal margin broadly and regularly rounded. Beaks at the anterior end and obtuse. Umbones prominent. Umbonal ridge angular and extending to the post-basal extremity. Valves somewhat flattened below the umbones and bearing an undefined shallow sulcus that becomes obsolete before reaching the basal margin. Post-cardinal slope flattened. Surface of the casts marked by faint concentric lines. We have four specimens belonging to this species, all of them are casts, and each one is broken at some point, but together, they show the entire valves. No part of the shell is preserved, and no muscular scars can be distinguished.

While this species has little resemblance to any other one in this genus, we have little doubt that it is a true *Cypricardella*.

Found by R. A. Blair, in the Choutean limestone, at Sedalia, Missouri, and now in the collection of S. A. Miller.

Family ORTHONOTIDÆ.

PALÆOSOLEN OCCIDENTALIS, n. sp.

Plate II, Fig. 13, cardinal view; Fig. 14, left valve of same specimen, the posterior part is broken away.

Species quite as large or larger than the type of the genus. Shell solenoid but the true length unknown. From other fragments than the one illustrated it is inferred that it represents only the anterior half and that the length is five times the height, but it may be only four times the height. Shell subcylindrical, basal and cardinal margins subparallel. Anterior end obtusely pointed. Posterior end unknown. Beaks near the anterior end, low, but curved over the cardinal line. Umbones rounded and fade away into the general convexity of the shell. Muscular impression large, round and anterior to the beaks. Ligament evidently external. Pallial line not observed. Hinge teeth, if any, unknown. Surface marked with distant concentric lines of growth which are crossed by a few, distant, radiating lines shown upon the cast. The shell, itself, is not preserved.

This species is readily distinguished from *P. siliquoideus*, the only other known species, by the general outline and surface characters, and yet, we think they are congeneric. The muscular impression is unknown in *P. siliquoideus*, but it must be quite small, while, in this species, it is very large. *P. siliquoideus* shows no evidence of an external hinge ligament, and Hall said that it was, probably, internal; but this species, we think, clearly had an external ligament.

Found by R. A. Blair, in the Choutean limestone, near Sedalia, Missouri, and now in the collection of S. A. Miller.

FAMILY CYTHERODONTIDÆ.

SCHIZODUS SEDALIENSIS, n. sp.

Plate II, Fig. 15, cardinal view; Fig. 16, right valve.

Species medium or below medium size. Shell subquadrate. Our specimens are casts, and somewhat compressed or only moderately convex. Length somewhat greater than height. Anterior end rounded. Basal margin broadly rounded. Posterior end subtruncate and gradually rounded into the basal margin. Cardinal line straight and prolonged posteriorly which produces the subquadrate outline to the shell. Beaks anterior to the middle of the shell, acute, and rising above the cardinal line where they are slightly incurved. Posterior umbonal slope somewhat angular and fading away toward the lower posterior part of the shell. Anterior umbonal slope rapidly merges into the depressed convexity or cuneiform shape of the shell. Pallial line regularly curves from one muscular scar to the other. Posterior end of the cardinal line subalate. Surface of the casts smooth and outline of the muscular scars not indicated.

This species is readily distinguished from all others, but is probably as nearly related to *S. medinensis* as to any other. But the shell of that species has a subtrigonal outline, while the outline of this species is subquadrate.

Found by R. A. Blair, in the Chouteau limestone, at Sedalia, Missouri, and now in the collection of S. A. Miller.

Remarks—We have found no species among the Lamellibranchs, from the Chouteau Group of Missouri, that occurs in any higher or lower Group of rocks. The genera, however, with the exception of *Blairella*, the new genus above described, are known to range geologically, as follows: *Posidonomya*, from the Upper Silurian to the Coal Measures; *Macrodon, Schizodus, Lunulicardium* and *Edmondia* from the Devonian to the Coal Measures; *Elymella* and *Sphenotus* from the Devonian to the Chouteau; *Cypricardella* from the Devonian to the Warsaw; *Mytilarca* from the Upper Silurian to the Chouteau. *Lioptoria* and *Palæosolen* were, heretofore, known only from Devonian rocks, and their range is now extended to the Chouteau. *Chænomya* was, heretofore, known only from the St. Louis Group to the Coal Measures and it is now brought down to the Chouteau. Species, heretofore described, from the

—3

Chouteau not referable to either of the above named genera, belong to *Pernopecten*, which ranges from the Chouteau to the Coal Measures, and *Grammysia*, which ranges from the Upper Silurian to the Chouteau.

The position of the Chouteau, at the base of the Subcarboniferous System and above the Devonian, is well established, by its crinoidal fauna; but the Lamellibranchs, as above set forth, furnish additional evidence of its place in the geological column, that cannot be misunderstood by any palæontologist. We have seen fragments, too poor for specific description, from the Chouteau, belonging to four other genera of Lamellibranchs, which further support the conclusion derived from those above described.

CLASS GASTROPODA.

ORDER BRANCHIFERA.

Family PLEUROTOMARIIDÆ.

MURCHISONIA INDIANENSIS, n. sp.

Plate II, Fig. 31, lateral view, part of the shell is preserved and a little of the surface ornamentation; Fig. 32, part of surface magnified.

Species very large. Shell elongated so as to be from one-fourth to one-third higher than wide. Volutions five or six. Only three volutions are preserved, in our specimen, but, at least, two are broken away. The last volution is sharply angular at the peripheral band, those toward the apex obtusely angular. The body whorl is rounded below, and slightly concave, from the suture to the peripheral angle. The concave depression is more strongly marked on the volutions toward the apex, between the suture and the periphery. Umbilicus open. Aperture subquadrate, about as high as wide, notched at the periphery. Suture canaliculate. Surface marked by fine striæ directed gently backward from the suture. No revolving ridges.

This species is readily distinguished by its large size, general form, angular whorls, subquadrate aperture, open umbilicus and surface ornamentation. We do not know of any nearly related species.

Found by Geo. K. Greene in the Knobstone Group, at New Albany, Indiana, and now in the collection of Wm. F. E. Gurley.

FAMILY CYCLONEMIDÆ

HOLOPEA GRANDIS, n. sp.

Plate II, Fig. 33, basal view, showing the form of the mouth and the open umbilicus; Fig. 34, lateral view, one whorl broken off at the apex.

Species very large. Shell about as high as wide. Volutions large, ventricose, and increasing rapidly from the apex. Four volutions, regularly rounded externally. Sutures sharply defined. Umbilicus open. Aperture rounded, subovate, somewhat flattened on the inner side, where it is in contact with the preceding volution. Surface ornamented with larger and smaller revolving striæ, which are crossed by finer oblique lines that cancellate the shell. On the body whorl, the largest revolving ridge is above the periphery, and the lines from the suture are directed obliquely backward to it, and then curve forward from it, and then backward over the lower rounded side of the volution.

This species cannot be classed with *Pleurotomaria*, because there is no notch in the aperture. It is not a *Cyclonema*, because it has an open umbilicus. It agrees with the generic characters ascribed to *Holopea*, and for that reason it is referred to that genus. There is no described species of *Holopea* so much resembling it, that any comparison is necessary.

Found by Geo. K. Greene in the Knobstone Group, ner New Albany, Indiana, and now in the collection of Wm. F. E. Gurley.

CYCLONEMA PULCHELLUM, n. sp.

Plate III, Fig. 6, lateral view of a specimen embeded in rock.

Shell large, elongate, subglobose-conical. Height greater than the width. Volutions six, the apex is broken off from the specimen illustrated. The volutions gradually expand from the apex to the last one, which rapidly enlarges to a very ventricose whorl. No umbilicus. The aperture is embedded in the rock, in our specimen, but it is apparently subcircular. Suture sharply defined not canaliculate. There are three strong revolving carinæ on each whorl, somewhat equally distant from each other and the suture, on the smaller volutions, and without diverging from each other or leaving the suture, they all occur, on the upper part of the ventricose whorl, above the periphery. There are a few smaller revolving carinæ that increase in number and spread over the last ventricose volution. There are numerous finer striæ directed gently backward from the suture that cross the revolving carinæ in sigmoid flexures and crenulate the carinæ so as to beautifully ornament the shell. The crenulations are hardly visible to the naked eye but are very distinct under an ordinary magnifier.

This species so far as disclosed, by our specimen, is a *Cyclonema*, possibly, the aperture may be different from a typical species, but we have no doubt, at present, of the generic reference. It is widely separated, however, by its general form and surface ornamentation from all other described species, and no comparison is necessary to distinguish it.

Found by Geo. K. Greene, in the Knobston Group, near New Albany, Indiana, and now in the collection of Wm. F. E. Gurley.

FAMILY EUOMPHALIDÆ.

STRAPAROLLUS MISSOURIENSIS, n. sp.

Plate II, Fig. 35, a specimen preserving the central whorls; Fig. 36, a large specimen, with central whorls broken.

Shell discoid; below medium size. Spire below the plane of succeeding volutions. Volutions five or six, lying in the same plane, slender and very slowly expanding. Transverse section of a whorl nearly circular, but probably slightly ovate with the narrower end on the inside of the volutions. The inner whorls appear to be round, but a subovate form is assumed as the aperture is approached. Aperture not preserved, in any of our specimens. None of the specimens collected exceed an inch in diameter. The surface is generally smooth, but the better preserved specimens show very fine transverse lines, under an ordinary magnifier.

We have numerous fragments of this species and some of them preserve the shell in excellent condition, with five transverse lines resembling those common on *Spirorbis*. The inner whorls might very readily be mistaken for a *Spirorbis*. It resembles *S. clymenioides*, from the Upper Helderberg Group, more than any other species known to the authors. It is distinguished by the more slender whorls, nearer circular transverse sections and surface ornamentation. It is also a smaller species.

Found by R. A. Blair, in the Chouteau limestone, at Sedalia, Missouri, and now in the collection of S. A. Miller.

FAMILY BELLEROPHONTIDÆ.

BELLEROPHON BLAIRI, n. sp.

Plate III, Fig. 7, dorsal view of a cast; Fig. 8, dorsal view showing the shell below the aperture.

Shell medium size, involuted, subglobose. Volutions expanded very moderately until the aperture is approached, when there is a more marked expansion, and recurved lateral lips. Umbilicus small, outer lip with only a moderate sinus in front and sides expanded, recurved and narrowly rounded. Aperture transverse and subreniform. The volutions are rounded toward the apex, but subangular on the sides of the last whorl. A slender dorsal band appears on the last volution which is somewhat obscure on the cast but well defined on the shell. The surface of the shell is ornamented with numerous fine striæ that arise from the dorsal band and curve gently forward and then backward over the obtusely subangular sides where they become obsolete. These transverse striæ are not visible on the cast.

This species is of the type of *Bellerophon bilobatus* from the Lower Silurian and there are several Silurian and Devonian species that bear a more or less striking resemblance to it. There is no difficulty, however, in distinguishing the species on making a comparison. There is no defined species from the Subcarboniferous rocks with which any comparison is necessary. The general form and surface ornamentation readily distinguishes it among species from rocks of the same geological age.

Found by R. A. Blair, in whose honor we have proposed the specific name, in the Chouteau limestone, near Sedalia, Missouri, and now in the collection of S. A. Miller.

BELLEROPHON SEDALIENSIS, n. sp.

Plate III, Fig. 9, dorsal view of a cast; Fig. 10, lateral view of part of a specimen preserving the shell.

Shell medium size, involute, subglobose. Volutions expanded very moderately until the aperture is approached when there is a marked lateral expansion. Umbilicus open. Outer lip with a moderate sinus and expanded narrowly rounded sides. Aperture transverse and subreniform. The volutions are rounded toward the apex, but subangular on the last whorl. The cast shows a central dorsal band, on the last volution, with a furrow on each side bordered by a sharply angular longitudinal line. The shell of this part of the last whorl is not preserved on any of our specimens. The shell is ornamented with numerous longitudinal, revolving furrows separated by fine angular striæ. These revolving furrows are visible upon many of the fragments of the casts, but much more strongly marked upon the fragments of the shell, wherever it is preserved.

This species is readily distinguished from *B. blairi* by the revolving furrows, and from all other described species by the general form and surface ornamentation. Species of *Bellerophon* have been described from the upper Taconic system and from nearly every recognized group of rocks up to the Upper Coal Measures. About ninety species have been illustrated and those which have been named and not illustrated might as well be struck out of the list, for they cannot be recognized by the definitions alone. There is such a general resemblance in the fossils belonging to this genus, that a common observer having learned one species can tell a *Bellerophon* wherever he sees it.

Found by R. A. Blair, in the Chouteau limestone, near Sedalia, Missouri, and now in the collection of S. A. Miller.

CLASS PTEROPODA. CLASS GASTROPODA.

ORDER CONULARIDA, n. ord.

This name is proposed to receive conical and pyramidal, pelagic shells, which may or may not have been contracted toward the mouth, but the texture of which is always horny with lime phosphate. The shells during the lives of the animals were flexible. They are smooth, or longitudinally divided and transversely furrowed. There are no muscular scars on the casts or on the shells. All belong to the palæozoic rocks. We refer to this order the family *Conulariidæ*, and the family *Enchostomidæ* hereinafter described.

Waagen used the word *Conularida* in Palaeontologica Indica, page 175, without defining it, or in any way limiting it, except to say, "they were certainly not pelagic shells," and to include in the order three families, which are widely distinct from each other, viz.: *Conulariidæ*, *Thecidæ* and *Tentaculitidæ*. He did not redescribe the family *Conulariidæ*, or describe any species belonging to it or in any way attempt to throw any light upon it. He had before him shells belonging to the family *Hyolithidæ* or *Thecidæ*, and they were the shells for which he was trying to provide a new Ordinal name, but they cannot be referred to *Conularida* for any reason that he suggested or for any other reason thus far put forth. We use the word *Conularida* for pelagic shells having lime-phosphate and not in the sense in which Waagen used it. It is the natural and proper Ordinal name to include the family *Conulariidæ* and cannot be preoccupied for any other purpose, besides, it has been used by others without defining it, for substantially the same purpose that we now use it. The family *Conulariidæ* does not belong to any living order of animals, and hence the necessity for providing an Ordinal name to receive it. And the same may be said of *Enchoslomidæ*.

Family CONULARIIDÆ.

There have been described, from the Palæozoic rocks of North America fifty-nine species of *Conularia*, besides three that have been named, but too poorly defined to be recognized. Among the fifty-nine species is *C. gattingeri*, which was named by Safford, in the Geology of Tennessee, p. 289, and compared with *C. trentonensis*, by saying, that it is larger and about ten inches long, and that it was found by Dr. Gattinger, while digging a cellar for his house, in the trenton limestone, in Nashville, Tennessee. This definition is exceedingly imperfect, but Dr. Gattinger made numerous plaster casts of the specimen and distributed them among the scientific men of the country (one of which was presented to S. A. Miller, by Dr. Gattinger, about twenty years ago), which made the form very generally known, and the remarkable size, if other specimens have been found, has prevented any synonym from being made. The species has been recognized, in all catalogues, and Dr. Gattinger has kindly loaned the original specimen, to S. A. Miller, for examination and description, and we propose to describe it, in order that the form may be better known and the specific name retained. We have an invariable rule to never name a new species without describing and illustrating it; but this is not our species, and as a single figure will cover half a plate, we will content ourselves by writing a description of the specimen.

The shell rapidly expands, is subquadrate in transverse section, but the diameter is greater in one direction than in the other. The sides are concave, in the superior part, which may or may not be the normal condition, because the shell is flexible in this genus. The four angles are deeply furrowed. There is a longitudinal line, in the middle of each side. The shell consists of an inner, black, horny layer and an outer, phosphatic layer. Where the outer layer is decorticated the surface is ornamented by transverse, arching furrows separated by narrow, smooth, elevated lines; but, where the outer layer is preserved, the furrows and ridges are about equal in width, and the ridges become crenulated costæ. The costæ are not regularly arched, but curve rather abruptly across the mesial line and are then directed, in nearly straight lines, inclined about ten degrees, to the furrows, at the angles. There are about forty-five transverse, crenulated costæ in an inch in length. The specimen near the larger end, where best preserved, has a diameter one way of two and two-tenths inches, and the other way of one and nine-tenths inches. It tapers, toward the apex, in a distance of three and six-tenths inches, and in the other of one and two-tenths inches. At the smaller end the specimen is broken off diagonally, and at the larger end an inch and a half in length of one of the wider sides is bent down as if approaching the mouth, but the other sides are continued without being bent and show the continuing enlargement of the shell. The greatest length of any part of the shell, that is preserved, is six inches. These measurements indicate that the specimen, when perfect, exceeded ten inches, in length. The surface ornamentation is altogether different from *C. trentonensis,* and the two species can never be mistaken for each other by any palæontologist.

Species of *Conularia* have been described from Trenton, Hudson River, Niagara, Lower Helderberg, Oriskany, Upper Helderberg, Marcellus Shale, Hamilton, Portage, Chouteau, Kinderhook, Waverly, Burlington, Keokuk, Warsaw and Kaskaskia Groups, and from the Lower and Upper Coal Measures. The range is from the early Trenton to the close of the Upper Coal Measures.

The shells are all pyramidal, and vary, in different species, from square and subquadrate, to octagonal and somewhat rounded. They expand slowly or rapidly in different species, and, so far as known, are contracted near the mouth. The mouth appears to have been very large, and no operculum or other shelly covering has ever been found belonging to it. We have examined more than one hundred specimens of *Conularia,* and have never seen the mouth of a single shell, so that what we have said about the mouth is on the authority of others. No muscular scar has ever been found inside the shell or on a cast, by which the animal was attached to the shell. The four angles of the shell are more or less furrowed, and a mesial line, on each side, is always indicated, and sometimes it amounts to a furrow. The shells are ornamented with transverse lines and furrows and costæ, some of which are smooth, others are crenulated, and all are more or less arched toward the mouth. The texture of the shell is horny, with lime-phosphate. The phosphate is conspicuous, in the outer layer. The phosphatic appearance is more strongly marked in some groups of rocks than in others, which is likewise true concerning the horny texture, which, sometime, as in *C. greenei*, resembles the test of a crustacean.

The genus *Conularia* is so distinct from all others that no other genus has ever been confounded with it. It is the only genus in the family *Conulariidæ*. Any one having ordinary perceptive faculties, after having carefully examined a specimen belonging to any species of *Conularia*, can tell a *Conularia* wherever he sees it, no matter to what species it belongs. This cannot be done with any other fossil specimen from the palæozoic rocks except, possibly, a *Bellerophon* or an *Orthoceras*.

The genus made its appearance, in the Trenton period, represented by small and large species, as fully developed and possessed of as distinctive specific characters, as the genus ever acquired. These it retained, throughout its life history, and closed its career in the Coal Measures, by such large species as *C. roeperi* and such small species as *C. crustula*. It came from some quarter wholly unknown, and after having lived as long as any other genus ever did upon the face of the earth, except, possibly, *Bellerophon*, *Pleurotomaria*, *Murchisonia*, *Orthoceras* and one or two genera of the *Brachiopoda*, it disappeared as abruptly as it came, without leaving a trace of its final course behind it. There is no evidence of development or evolution connected with the genus. It never showed any higher or lower stage of existence, than it did when it first appeared. Some species had a wide geographical and geological range; for example, *C. trentonensis* from New York and Kentucky, and *C. subcarbonaria* from the Keokuk Group, at Crawfordsville, Indiana, and Hamilton, Illinois. We have seen large specimens and small specimens belonging to the same species, possessing exactly the same ornamentation and surface characters. But we have never seen anything that indicated advancement or decline in the genus or in any species, and further, we have never seen any intermediate forms, that might be said to represent a link connecting any two species. This may be cold comfort, to those limited palæo-biologists, who claim to see, in every fossil, a link from the lowest to the highest stages of animal existence. It is, nevertheless, true, that we do not even know to what Class, in the animal kingdom, the family *Conulariidæ*, or the Order, *Conularida* should be referred.

—4

The family *Conulariidæ* has been dumped into the Class *Pteropoda*, by some authors, and into the Class *Gastropoda*, by others, and, probably, the reason has been about as good in the one case as in the other; for it may have no near relation to either. It is like the *Graptolitidæ*, no one knows to what Class it belongs, though every author, having anything to do with the family, will drop it into some Class, and say nothing about the reason for doing so.

It is not scientific to name a Class, in the subkingdom Mollusca, when the definition of the Class and a single order belonging to it must, necessarily, be the same. We have gone as far, in Classification, by briefly defining a new Order, as it is practicable to go, in the present state of the learning, besides, we anticipate future discoveries will throw important light upon the subject. *Conularia* belonged to the great ocean, but whether its home was in the depths or near the surface, we do not know. If it had been a littoral shell, it would not have found a place in so many Groups of rocks, where other littoral shells are unknown. It possessed a hardy shell, capable of preservation in nearly all deposits, or we would not find it, with its peculiar purple, phosphatic lustre, in sandstones, clays, shales, and limestones. We find it scattered here and there, generally very rare, and never in abundance, which indicates that we have not found it in its best and favored habitations. When we find it in abundance, we may find and recognize its relatives, and, even before that time, or at any time, we are liable to see them unearthed, for we have seen only a very small part of the rocks belonging to our country.

CONULARIA ROEPERI, n. sp.

Plate III, Fig. 1, middle part of a specimen, wider side; Fig. 2, transverse section.

Species large, long, slowly expanding, pyramidal, subquadrate, in transverse section. Transverse diameter a little greater between the sides one way than the other. Sides slightly convex. The four angles deeply furrowed. Longitudinal line in the middle of each of the four sides, but it cannot be called a furrow, as it does not, in all cases, break the transverse costæ. Surface ornamented with transverse arches, shallow, smooth furrows, that are separated by fine lines or costæ. In passing across the sides the costæ curve forward toward the aperture, and sometimes alternate in the middle, and at other times cross the mesial line almost without interruption. The distance between these transverse lines does not seem to increase with the size of the shell, but is uniform throughout the length of our specimen. The lines do not curve forward toward the aperture in the furrows, at the angles, as is usual in this genus, but they arise from the bottom of the furrows and cross the sides in regular arches. There are about fifty of these transverse lines in an arch.

Our specimen has a length of nearly five inches. The transverse diameter at the small end is an inch one way, and an inch and one-twentieth the other, and at the large end an inch and a half one way and an inch and six-tenths the other. It is quite evident if the specimen was complete it would be more than a foot in length.

It is unnecessary to compare it with any other specimen for the purpose of distinguishing it.

Found in the Coal Measures of Luzerne County, Penn., and presented to S. A. Miller by Rev. Wm. Roeper, an ardent collector and naturalist, in whose honor we have proposed the specific name.

CONULARIA GREENEI, n. sp.

Plate III, Fig. 3, middle part of a specimen.

Species long, slowly expanding, pyramidal, subquadrate, sides equal, deeply furrowed at the four angles. Longitudinal line in the middle of each side. Surface ornamented with wide, concave, smooth furrows that arch forward from the four angles. These furrows are separated by sharp costæ generally without crenulations. The costæ sometimes cross the mesial line without interruption, in other cases they terminate alternately at the mesial line. They do not curve forward when bending into the furrows at the four angles, nor do they reach the bottom of the furrows. They alternate in these furrows. The transverse furrows are crossed by a few longitudinal wrinkles, at the margin of the longitudinal furrows.

There are only thirteen transverse furrows in an inch, where our specimen is eight-tenths of an inch in diameter. The shell of our specimen is horny, and has the smooth, hard appearance of the test of a trilobite.

This species is so different from all that have heretofore been described, from the Keokuk Group, that no comparison with any of them is necessary. It is distinguished by its slender form, wide, transverse, smooth furrows and sharp costæ. There are some slight crenulations on the costæ, toward the larger end of our specimens, but none toward the smaller end.

Found by Geo. K. Greene, in whose honor the specific name is given, in the Keokuk Group, at Edwardsville, Indiana, and now in the collection of Wm. F. E. Gurley.

CONULARIA SEDALIENSIS, n. sp.

Plate III, Fig. 4, fragment from the middle part of a specimen, somewhat twisted; Fig. 5, under side of the shell showing the nodes on the costæ.

Species large, rather rapidly tapering, pyramidal, subquadrate, in transverse section. We have several fragments of this species that are twisted and curved in different directions, showing the great flexibility of the shell itself. The specimen illustrated in Fig. 4, presents the largest undisturbed surface of any of them. The sides, as near as can be determined, are flat and equal. The angles are not very deeply furrowed, and the longitudinal line in the middle of each side is not very strongly marked. Surface ornamented with wide, transverse, arching furrows that are separated by coarse costæ. In passing across the sides, the costæ curves forward toward the aperture, and sometimes alternate at the mesial line, and at other times cross it without apparent interruption. The costæ are geniculated at the furrows, at the four angles. The inner layer of the shell is of a light gray color and differs very little in color from the limestone matrix; the second or middle layer is of a reddish brown color and horny texture. The costæ bear a line of strong tubercles which are so fixed in the matrix that the shell is split and decorticated in removing it from the matrix. The tubercles and the middle layer of the shell are broken away from the costæ on the specimen illustrated by Fig. 4, but the second layer is preserved in many of the furrows, where it is perfectly smooth. In some places, on some of the specimens, the tubercles may be seen on the costæ, but they are best shown in the matrix after the shell is taken out, as shown by figure 5. There are thirteen costæ to an inch in length, where a side is one and one-third inches wide, and forty-two tubercles on one of the costæ in an inch in length.

This species is distinguished by its wide transverse furrows, coarse costæ and strong, distant tubercles, without other ornamentation.

Found by R. A. Blair, in the Burlington Group, at Sedalia, Missouri, and now in the collection of S. A. Miller.

ENCHOSTOMA n. gen.

[Ety. *enchos*, sword; *stoma*, blade.]

Shell smooth, elongate, lanceolate, transverse section more or less rounded or narrowly subovate. Shell substance thin, solid, flexible, horny, lime-phosphate. Type *Enchostoma lanceolatum*, described as *Hyolithes lanceolatus*, S. A. Miller, 1892, advance sheets, 18th Rep. Geo. Sur. Ind. p. 63, from the Chouteau limestone.

When the species *Hyolithes lanceolatus* was described, only a few fragments had been selected and the specimen then illustrated was supposed to represent nearly the complete length, but later collections from Sedalia and Providence, Missouri, showed it was not half the length of the original. We have a specimen now before us two inches in length, which is broken off at both ends, and the smaller end is as large as the smaller end of the type, which is less than an inch in length. Another specimen at hand an inch and a quarter in length, is larger at the smaller end than the type is at the larger end. Another fragment an inch in length is no larger at the larger end than the type is at the smaller end. The evidence thus furnished shows that a complete specimen would be three inches in length or even more than that, and that the greater diameter at the larger end, is three-tenths of an inch. We have examined about fifty fragments, none of them seem to be complete at either end, but as none of them seem to contract toward the larger end, we infer that the species does not contract toward the mouth, as in *Conularia*. The smaller end of all our specimens, though in some cases, not exceeding one twenty-fifth of an inch in diameter, is broken off, so that evidently a perfect specimen has an acute point. All of the specimens from the apical end of the shell show a curvature, and the best specimens show a curvature of an eighth or tenth of a circle. The apical end is round in the best preserved specimens, but all of them are subovate toward the mouth. As many of them are compressed toward the larger end it is hard to tell the correct transverse section, but from the large number examined, it is clear that the section illustrated in fig. 36, pl. IX, of the 18th Rep. of the Geo. Sur. Ind., is somewhat compressed. A normal section, probably, becomes more and more acutely ovate as the mouth is approached.

A few fragments on hand are longitudinally fluted, but if they are normal, they belong to a distinct and undescribed species.

The texture of the shell distinguishes this genus from the *Hyolithadœ* and brings it into some kind of relation to *Conularia*. It cannot, however, be fairly classed with the *Conulariidœ* though properly falling within the order *Conularida*. We, therefore, propose the family name *Enchostomidœ*, but the family name as at present understood, will take the same definition that the genus has. We have seen fragments of a long, arching, round shell, somewhat, in form, like a *Dentalium*, in limestone, belonging to the Keokuk Group, but having the shell texture of this genus, that may be generically distinct and if so the family *Enchostomidœ* may be defined and limited so as to include two genera.

The shells in the families *Hyolithidœ* and *Tentaculitidœ* are thick and composed of layers that may sometimes be horny, but they are never phosphatic. There is as much difference in the texture of the shells of *Conularia* or *Enchostoma* and *Hyolithes* or *Tentaculites* as there is between the shells of *Lingula* or *Discina* and *Orthis* or *Spirifera*. And there is as much reason for placing *Conulariidœ* in an Order distinct from *Hyolithidœ* and *Tentaculitidœ* as there is for dividing the *Brachiopoda* into the Orders *Lyopomata* and *Arthropomata*. The fundamental difference in the composition and texture of the shells is the basis of the separation into Orders. The general form of the shells in the genera *Conularia*, *Enchostoma*, *Hyolithes* and *Tentaculites* is altogether different as well as the composition and texture. *Conularia* are pyramidal *Enchostoma* round and curved toward the apex and ovate toward the mouth, *Hyolithes* short, flattened on one side and straight, and *Tentaculites* straight, round and annular.

As a general rule a palæontologist is able to classify the fossils with reference to some known living organism. He finds a trace or path from the unknown animal to the known, and reasons forward from remote ages to the present, and he finds here and there a fauna that characterizes a geological age and enables him to determine it at distant localities, but the *Conularida* at present are to be classified with the unknown, save that they are evidently mollusks and belong to the great Palæozoic ages.

CLASS CEPHALOPODA.

ORDER TETRABRANCHIATA.

FAMILY CYRTOCERATIDÆ.

CYRTOCERAS DUNLEITHENSIS, n. sp.

Plate III, Fig. 11, lateral view, showing a great part of the chamber of habitation; Fig. 12, transverse section.

Shell medium size, strongly curved and regularly enlarging from the apex to the mouth. The siphuncle is on the ventral side or outer margin of the curve and produces an expansion of the shell, which forms, as shown in a transverse section, the narrow end of a sharply ovate figure, at the ventral margin. The sharply ovate transverse section is represented in figure 12. Twelve septa have a length, on the inner curve or dorsal side, of six-tenths of an inch, and, on the ventral side of one and eight-tenths inches, where the dorso-ventral diameter, at the smaller end, is half an inch, and, at the larger end, nine-tenths of an inch. The greatest lateral diameter is about the dorsal third of the shell where it measures, at the same sections, thirty-five hundredths of an inch, at the smaller end, and eighty-five hundredths of an inch, at the larger end. The septa cannot be distinguished near the apical end of our specimen. The chamber of habitation is, probably, nearly complete, in our specimens, and it constitutes more than one-third of the entire length of the shell. Part of the shell is preserved on the inner dorsal side and shows regular lamellose lines of growth, without other ornamentation. Part of the shell is also preserved over the siphuncle and posterior ventral side of the body chamber, which shows the lamellose lines of growth curve backward, in crossing the siphuncle, and indicates a notch or sinus at the ventral lip of the aperture. Where the shell is decorticated, the cast is smooth.

This species is distinguished by its strong curvature, rapid enlargement, sharply ovate transverse section, long body chamber and lamellose lines of growth.

Found in the Trenton Group, at Dunleith, Illinois, and now in the collection of Wm. F. E. Gurley.

FAMILY ORTHOCERATIDÆ.

ORTHOCERAS CALDWELLENSIS, n. sp.

Plate IV, Fig. 1, middle part of a specimen; Fig. 2, transverse section.

Shell straight, large, long, very slowly and regularly enlarging from the apex toward the mouth of the chamber of habitation. Only the middle part of the shell is preserved in our specimens. Chamber of habitation unknown. Transverse section subelliptical. Siphuncle subcentral. The shell is preserved, on our specimen, and the air chambers are not, therefore, exposed. The shell is widely and deeply annulated or transversely furrowed. The dividing ridges are nodose. The nodes are arranged in longitudinal rows. There are fourteen nodes on each transverse ridge, in the specimen, and hence, there are fourteen longitudinal rows of nodes. A longitudinal line crosses each furrow from node to node, but it is nearly obsolete at the bottom of the furrows. The width of a furrow or distance between two nodes, at the larger end is equal to one-third of the shorter diameter of the shell; but, at the smaller end of the specimen, the distance between two nodes is more than one-third of the greater diameter. The width of the annulatinus, therefore, do not bear a regular proportion to the diameter of the shell. There is an obscure node between the regular nodes, at the larger end, but none near the smaller end. The septum shown, at the smaller end, is highly arched, and, it appears as if there is only one septum to correspond with each annulation. The shell is thick, and the outer surface of the furrows shows no lamellose lines of growth, but, possibly, a better preserved specimen would show such lines.

This species has more resemblance to *O. nodocostum*, from the Niagara Group, than to any other described species. *O. nodocostum* is frequently classed as a synonym for *O. annulatum*, but the two species are distinct and are readily distinguished, by any palæontologist, from an examination of the shells or the casts. The annulations are wider and the nodes are not as prominent, in this species as they are in *O. nodocostum*, and the septa are evidently more distant from each other, and more highly arched. *Orthoceras* was a pelagic shell and, probably, lived as long upon the face of the earth as any other genus ever did. The annulated forms made their appearance, in the Lower Silurian age, and are found in all Groups of rocks, from there into the Subcarboniferous. The form called *O. annulatum* occurs, in the Niagara Group, on both sides of the Atlantic, and almost everywhere, that the rocks of that age are known to exist.

Found by James G. Caldwell, in whose honor the specific name is given, in the Upper Helderberg Group, in Clarke County, Indiana, and is now in the collection of Wm. F. E. Gurley.

FAMILY GOMPHOCERATIDÆ.

POTERIOCERAS JERSEYENSE, n. sp.

Plate IV, Fig. 3, side view of a specimen somewhat compressed.

Shell large, acutely obovate or balloon-shaped. Body chamber much longer than the septate portion. Greater diameter about the middle of the body chamber. Section subcircular. Our specimen is somewhat compressed so that a transverse section cannot be accurately determined. The body chamber appears to be bulged on one side. Probably, most tumid on the lower ventral side. Our specimen shows six air chambers and probably there were never more than two or three more. If complete, therefore, there would not be more than eight or nine small air chambers.

When compared with *Poterioceres missouriense* which this species most resembles, it will be observed, that the body chamber is one-half longer, and the septate portion much shorter, in this species, than it is in that one. The inclination of the septa or obliquity toward the tumid side is the same in both species.

Found by the late Wm. McAdams, in the Kinderhook Group, in Jersey county, Illinois, and now in the collection of Wm. F. E. Gurley.

FAMILY GONIATITIDÆ.

No one has described a *Goniatite* from the Lower or Upper Silurian rocks of America. The species described from the earliest rocks is *Goniatites mithrax*, from the Upper Helderberg Group, in Ohio. It is possible that the reference of this species, by the collector, to the Upper Helderberg was erroneous, because rocks of the age of the Hamilton Group, in Ohio, have been frequently referred to the Upper Helderberg, but we think that is not probable, and we have no right to assume such to be the case, without some evidence to support the assumption, and we have none. We only know that many species occur in the Hamilton Group, and this is the only one referred to older rocks. Where are its ancestors or from whence did it come?

It is a very large species, with four or more volutions. The outer one embraces the inner ones and closes the umbilicus. A transverse section of a volution is semi-elliptical, the dorso-ventral and transverse diameter being about as two to one.

"The septa curve gently forward, from the umbilicus for nearly two-thirds of the width of the volution; thence more abruptly backward, forming a broad, low, undefined saddle, to a point nearly three-fourths of the width of the volution, when they again bend forward to the margin of the periphery, leaving a broad, deep lobe, which occupies nearly one-third the width of the volution; and thence turning abruptly backward to near the center of the periphery, and sharply recurring, leave an acute triangular saddle on each of the margins, and a narrow, acute,

—5

ventral lobe. The saddle occupying the center of the short, ventro-lateral curve is acute at the summit, having a height one-fourth greater than the width at the base, and curving a little more abruptly on the ventral side. The ventral lobe extends about half the depth of the adjacent air-chamber, and is abruptly narrowed below, the walls being essentially parallel and coïncident with those of the siphuncle. The septa are thin in the center, thickened and imbricating at the margins, leaving a deeply marked suture line. "

The definition is from Hall's Palæontology of New York, vol. 5, pt. 2, p. 433, and it will be observed, that it includes all the characters ascribed to the genus. If the large size, involute whorls and complicated chambers do not indicate a fully developed *Goniatite*, we would like to know what later species took on characters belonging to a higher stage of animal development. We know there are some more recent species having more angles in the septa, and others with fewer angles, but the increase or decrease, in the number of these, will hardly be held to indicate a higher or lower stage of the development of the animal; for, if so, we need only to turn our attention to angles in the septa, to rate the species in the grade of its animal existence. If the stage of the involution is the measure of the perfection of the animal, then this species reached the highest grade, for the outer volution embraces all the inner ones, and, we cannot assume the contrary, because no older cephalopod ever embraced the inner whorls, in the outer volution, and closed the umbilicus.

What these facts tend to prove is that, so far as we know, the most ancient specimen belonging to the family *Goniatitidæ* that has ever been found in America was as highly developed as an animal and in the structure of its shell, as any more recent specimen. Prof. Hyatt raises the *Tetrabranchiata* to the grade of a subclass and divides it into the orders, *Nautiloidea* and *Ammonoidea* and refers the family *Goniatitidæ* to the Order *Ammonoidea*. This is probably a good classification, but it does not alter the conclusions to be drawn from the facts we are presenting. His idea, however, that "The *efforts* of the *Orthoceratite* to adapt itself fully to the requirements of a mixed habitat of swimming and crawling gave rise to the *Nautiloidea*, and the *efforts* of the same type to become completely a littoral crawler evolved the *Ammonoidea*," does not meet with any support from the shells that have been discovered belonging to the *Goniatitidæ*. One can imagine that from *Orthoceras* through *Cyrtoceras, Gyroceras* and other genera arose the *Nautilidæ*, but there is absolutely no connecting link between the *Orthoceratite* and the *Goniatite* that has ever been discovered, and consequently no mental conception can be introduced to supply the omission. We describe below twelve species, eleven of which are new, and refer them to the genus *Goniatites*. They are from different Groups, in the subcarboniferous system and from the Coal Measures. Some have a closed umbilicus, others an open one, the volutions differ greatly, in form, and there are great variations, in the septa, some reversing the order of the sinuosites, in crossing the volutions; but we are unable to distinguish characters which we can call generic and by so doing separate them into different genera. We think Prof. Hyatt would not refer all the species to the same genus, and we appreciate his learning, but are unable to follow him in his generic subdivisions.

GONIATITES BLAIRI, n. sp.

Plate IV, Fig. 4, side view, part of it covered with the matrix; Fig. 5, portion of the ventral or outer margin.

Shell below medium size, discoid. The sides of the volutions are flattened and inclined toward the ventral or outer margin, which is narrowly rounded or subangular. The umbilicus is open and exposes part of the volutions. Our specimen shows less than one and a half volutions and it would appear that about half of each volution is exposed and that a complete specimen would contain about three volutions. Probably the last volution embraces less of the preceding one than the inner volutions do. The sides of the umbilicus are subangular and the greatest lateral diameter of a volution is near the umbilicus, or adjoining the abrupt descent to the umbilical cavity. The dorso-ventral diameter of a volution is very little more than the greatest lateral diameter. The volutions enlarge, at first, very gradually, but the enlargement is increased, toward the outer part of the last volution. The surface is marked by furrows, that are directed moderately backward, from the angle at the umbilicus, for about half the diameter of the volution, and then more rapidly curve backward to the periphery which is crossed by a rather sharp angle. The furrows are separated by sharp angular ridges. Septa and body chamber unknown.

We think there can be no doubt that this species belongs to *Goniatites*, though none of the septa can be seen. The species is distinguished by its general form, flattened volutions, subangular periphery and transverse curving furrows that form an angle on the ventral margin.

Found by S. A. Miller, in the Chouteau limestone, six miles from Sedalia, Missouri, and now in his collection. The specific name is in honor of R. A. Blair.

GONIATITES PARRISHI, n. sp.

Plate IV, Fig. 6, surface form of a septum; Fig. 7, lateral view; Fig. 8, ventral view.

Shell below medium size, discoid, sublenticular, volutions very rapidly expand. Transverse section of a volution semielleptical. The sides of the volutions are somewhat flattened and inclined toward the ventral margin, which is narrowly rounded. The outer volution embraces the inner ones. The umbilicus is small and discloses no part of the inner volutions. The sides of the umbilicus are abrupt and the greatest transverse diameter of a volution is near the abrupt descent to the umbilical cavity. The dorso-ventral diameter of a volution, in the early growth of the shell, does not exceed the transverse, but the dorso-ventral diameter increases more rapidly than the transverse, with the growth of the shell, and soon becomes one-fourth greater, and, probably in older shells than ours, it may become one-half greater. The external shell is unknown. The air chambers are very short and do not increase, in length, in proportion to the increasing size of the volutions. The septa are closely arranged.

Each septum curves gently from the umbilicus forward and back to near the middle of each lateral side, where it forms an obtuse retral angle and is directed nearly straight forward to the ventro lateral margin, where it makes an abrupt retral bend and is directed backward nearly to a line with the first formed angle, and then again bends forward and makes a forward semicircular curve across the median line of the ventral margin. There are, therefore, five saddles and five lobes, which will be best understood by looking at the illustration. The middle saddle curves forward slightly more than the lateral saddles, but the ventro-lateral saddles extend more than twice as far forward and are obtusely rounded at the anterior ends.

This species is distinguished by the general form of the shell and by the surface form of the septa.

Found by W. J. Parrish, in whose honor the specific name is given, in the Upper Coal Measures, at Kansas City, Missouri, and now in the collection of Wm. F. E. Gurley.

GONIATITES ELKHOBNENSIS, n. sp.

Plate IV, Fig. 9, lateral view; Fig. 10, ventral view; Fig. 11, surface form of a septum.

Shell very large, discoid. There are between three and four volutions preserved, in our specimen, and, apparently, an entire individual would have seven or eight volutions, all of which are exposed, in the very wide shallow umbilical cavity or depression. The volutions are rolled, in the same plane, and increase more rapidly, in transverse, than in the dorso-ventral diameter. At first the transverse diameter is not greater than the dorso-ventral, but later, as shown by the ventral view in figure 10, the transverse diameter becomes more than twice as great as the dorso-ventral. The ventral side is slightly convex and the dorsal side correspondingly concave, which allows the volutions to be very closely coiled, without properly overlapping. The inner volutions are beveled, on the lateral sides, from the middle part, leaving a middle angular ridge, which gradually approaches, in the last volutions, the ventral margin, and, at the body chamber, forms an angle, at the ventral margin, from which the beveled edge extends to the next inner volution.

The external shell of our specimen is not preserved. The air chambers are very long, but do not increase, in length, in proportion to the increasing size of the volutions, but, on the contrary, do not seem to increase in length at all. Some of the septa are not distinct toward the end of the last volution, and it is not clear whether or not any part of the body chamber is preserved, but the ventral side indicates that another volution is necessary to complete the shell. The septa are distant. They curve gently backward from the umbilicus and then forward each one crossing the lateral side of a volution in a sigmoid flexure, and are then more abruptly directed backward over about one-third the width of the ventral margin, where each one is abruptly bent forward and forms a semielliptical arch across the middle of the ventral side. There are, therefore, three convex saddles, the middle one being semi-elliptical, and extending only about half as far forward as the less convex lateral saddles do. This will be better understood by looking at the ventral view and the surface form of a septum as represented in the illustrations.

This species is distinguished by the general form of the shell, manner of enrollment of the volutions, transverse section of a volution and by the saddles and lobes in the septa.

Found by Geo. K. Greene in the Coal Measures, on Elkhorn
Creek, in Kentucky, and now in the collection of Wm. F. E.
Gurley.

GONIATITES MONTGOMERYENSIS, n. sp.

Plate IV, Fig. 12, lateral view; Fig. 13, ventral view; Fig. 14, surface form of a septum.

Species below medium size, globose, volutions slowly expanding.
There are between · three and four volutions preserved, in our
specimen, and, apparently, an entire shell would have more than
six volutions; part of the dorsal margin of each is exposed in the
deep funnel-shaped umbilicus. The volutions are rolled in the
same plane, and increase much more rapidly in the transverse,
than in the dorso-ventral diameter. At first, the transverse
diameter is not more than twice as much as the dorso-ventral,
but, at the end of our specimen, which is somewhere in the fourth
volution, the transverse diameter is three and a half times the
dorso-ventral and doubtless the end of the volution of an entire
shell has a transverse diameter five or six times as great as the
dorso-ventral. The ventral side is broadly convex and the dorsal
side correspondingly concave, for the width of the inner volution,
and between that and the margin, the outer volution is beveled to
the form of the funnel-shaped umbilicus. The lateral side of a
volution consists of a sharp denticulated edge. The umbilicus is
like a hollow cone or funnel bordered by a sharp denticulated
margin.

Six furrows arise, at the margin of the umbilicus, at depressions between the denticulations, and are directed forward at an
angle of about forty-five degrees, across one-fourth of the width
of the ventral side, and then cross the middle part of the ventral
side in a slightly undulating line. These furrows divide a volution
into six subequal parts, though the distance between them is not
uniform and does not increase regularly with the growth of the
shell. They cross the shell without any reference to the septa or
chambers. The air chambers are not of equal length, but they do
not increase in proportion to the increasing size of the shell. The
septa cross the ventral side in transverse waving lines. A septum
curves from the umbilicus forward and back in the form of a half
circle and then forms a retral subangular bend and again curves
forward and back in the form of a half ellipse, and again forms
a retral subangular bend and curves forward over the middle
part of the ventral side somewhat in the form of a half circle depressed in the middle part so as to make it bifid. There are,

therefore, five saddles and as many lobes. Four of the saddles are somewhat evenly convex, the middle one is only about two-thirds as high as those adjoining and is abruptly depressed, in the middle part, so as to make it bifid, and form a short narrow lobe in the middle of the ventral side. The illustrations will, at once, give a better idea of the septa than any definition can give.

This species is quite peculiar and is distinguished by the general form of the shell, by the hollow cone-like umbilicus surrounded by the sharp denticulated margin of the last volution, by the six furrows that cross the ventral side of each volution and by the saddles and lobes in the septa.

Found by the late Wm. McAdams in the Coal Measures of Montgomery county, Illinois, and now in the collection of Wm. F. E. Gurley.

GONIATITES FULTONENSIS, n. sp.

*Plate IV, Fig. 15, lateral view; Fig. 16, ventral view; Fig. 17,
surface form of a septum.*

Species medium size, subglobose, periphery regularly rounded; volutions rather rapidly expanding. Transverse section of a volution semi-elliptical, the transverse diameter being a little more than the dorso-ventral. Number of volutions not known. The last volution embraces all the inner ones. Umbilicus small, open but not disclosing the inner volutions. The sides of the volutions are slightly flattened and inclined toward the regularly rounded periphery. The sides of the umbilicus are abrupt, and the greatest transverse diameter of a volution is near the abrupt descent to the umbilical cavity. The external shell of our specimen is not preserved.

The air chambers are very complicated, of moderate length and do not increase in length with the increasing size of the volutions. The septa are close in some places and distant in others, depending upon the peculiar sinuosities. Within the umbilicus there is an angle in each septum in the overlapping part of each outer volution, from which the septum, in a gentle arch, turns over the margin of the umbilicus, and from an obtuse angle, curves forward and back, turning more than a half circle and extending back to an acute and prolonged point from which it takes a retral course and again curves forward beyond the first semi-circular curve, and then back to another prolonged point, where it takes another retral course and again curves forward beyond the second prolonged curve and then back to another acute and prolonged point, in line with the two preceding acute points, where it again takes a retral course and extends forward as far as the second prolonged curve, and instead of arching over the middle of the periphery, abruptly curves back a short distance and then forward and back so as to leave the summit of this saddle bifid, and to form a short narrow lobe in the middle of the ventral side. There are, therefore, seven saddles and seven lobes without including the small narrow lobe in the middle of the ventral side. The three saddles on each side rapidly increase in length from the umbilicus toward the periphery, and the one on the periphery has a length about equal to the middle one on each lateral side. The numerous sinuosities of the septa are best understood by observing the illustrations. When we look at an end view of a volution or at the face of a chamber, we see, not only the seven projecting saddles, the middle one of which is bifid, as above described, but also two short ones on each side at the mouth of the umbilicus, one of which is indicated by the gentle arch which turns over the margin above mentioned, and the other is within the mouth of the umbilicus and forms the inner angle of the truncated horn of the subcrescentiform chamber, and precedes the angle first above mentioned. This inner saddle is not disclosed, in a coiled shell, but the one on the margin of each umbilicus might very well be counted, making nine saddles in a septa. The siphuncle is rather large, and as usual, near the dorsal side

This species will be distinguished by its general form, the great number of sinuosities in the septa, and by the nine exposed saddles in each septum.

Found by John Wolf, in the Coal Measures, in Fulton county, Illinois, and now in the collection of Wm. F. E. Gurley.

GONIATITES KENTUCKIENSIS, S. A. Miller.

Plate V, Fig. 1, lateral view of a large specimen preserving the outer shell.

This species was described and illustrated in North American Geology and Palaeontology, page 439, from the inner whorls of specimens, that did not preserve any of the outer shell. The volutions are unusually numerous in this species, though the number in a mature shell is still unknown. Evidently there are ten or more. In a large shell two or three of the volutions may be seen by looking into the umbilicus, but in the younger specimens the inner volutions are not disclosed. The surface of the shell is covered with numerous, sharp, elevated, revolving lines, separated by wider revolving furrows.

In describing this species originally the other side of the volutions was called the dorsal side of the shell, following the terminology of Meek and most of the early authors; but Owen long since showed that in the living Nautilus, the ventral side of the animal is upon the outside and the dorsal side on the inner side of the volution, and most late authors have made their descriptions of the shells of Cephalopods conform to the position of the animal in the shell of the Nautilus. We have adopted this method, and the reader, in order to make comparison with the description by Meek of coiled Cephalopods in the Geological Survey of Illinois, and by other authors in North American Geology and Palæontology, and elsewhere, will find it necessary to reverse the words dorsal and ventral as applied to the shells, so that they may apply to the supposed position of the animal when within the shell, as evidenced by the position of the Nautilus.

The specimen here illustrated was found by Geo. K. Greene, at the typical locality, in the St. Louis Group, at Crab Orchard, Kentucky, and is now in the collection of Wm. F. E. Gurley.

GONIATITES LUNATUS, n. sp.

Plate V, Fig. 2, lateral view; Fig. 3, end of a volution and
ventral view, showing the surface markings on the shell;
Fig. 4, surface form of a septum taken from
thinner and smaller specimens; Fig 5,
end and ventral view of same.

Species large, globose, volutions rather rapidly enlarging and the periphery becoming more and more broadly rounded with the growth of the shell. Figure 2 is a lateral view of a large specimen, though incomplete. It preserves part of the outer shell and does not expose the septa. Figure 3 is a smaller specimen, showing the outer shell but none of the septa. Figure 4 represents a septum from a still smaller specimen, a ventral view of which is represented by figure 5. Number of volutions not known. A transverse section of a volution is lunate or crescentiform. The last volution embraces all the inner ones and closes the umbilicus. The air chambers are short and some parts of the septa come very close together.

—6

Each septum arises from the umbilicus and makes a broad curve backward beyond the middle of the side, where it makes a sharp retral angle and then curves forward and backward forming more than half an ellipse, and, instead of arching over the middle of the periphery, abruptly curves forward a short distance, and then backward and forward so as to make this lobe bifid, and to form a short narrow saddle in the middle of the ventral side. There are, therefore, five saddles and four lobes, but the middle saddle and two middle lobes are very small. The course of a septum is best understood by looking at the illustration. It will be observed that the courses of the septa, in this species, are the reverse of those in the species above described. That is, to form the first lobes, they are directed backward in this species, and forward in those above described. The bifid saddle is directed forward, in the species above described, and, in this, the curve crossing the middle of the ventral side is directed backward, and we have a small central saddle instead of a small central lobe, etc. The surface of the shell is marked by fine, transverse, imbricating lines of growth.

This species is distinguished by its general form, transverse lunate section of the volutions, fine transverse lines of surface ornamentation, and peculiar sinuosities of the septa.

Found by Geo. K. Greene in the Coal Measures, on Elkhorn Creek, in Kentucky, and now in the collection of Wm. F. E. Gurley.

GONIATITES ILLINOISENSIS, n. sp.

Plate V, Fig. 6, lateral view; Fig. 7, ventral view; Fig. 8, surface form of a septum.

Species medium size, subglobose, volutions moderately enlarging' and periphery broadly rounded. Our specimen exposes part of three volutions, leaving the impression that a complete shell contains not less than six volutions. A transverse section of a volution is subcrescentiform, the horns being short and obtuse. The last volution encloses all the inner ones, but leaves a rather large open umbilicus. The air chambers are short and complicated. The outer shell is not preserved in our specimen.

Each septum may be seen to curve backward across the obtuse end of the horn of the crescent, within the cavity of the umbilicus, and form an acute angle at the mouth of the umbilicus, where it curves forward and then backward, in a waving line, to an acute point, which is posterior to the first angle, where it makes a sharp retral angle and curves forward, in a waving line, anterior to the first forward curve or saddle and then backward, in a waving line, to an acute point, which is slightly anterior to the second one, where it makes another sharp retral angle and again curves forward ,in a waving line, to a level with the anterior part of the first saddle, and then abruptly curves back a short distance and then forward and back, so as to make the summit of this saddle bifid and to form a short, narrow lobe at the periphery, in the middle of the ventral side. There are, therefore, two saddles upon each side of the volution, and a bifid saddle in the middle of the ventral part, and one on each side of it, the latter being the longer ones. The two saddles, on the sides of the ventral margin, extend somewhat anterior to the others. The sinuosities of the septa are best understood by looking at the illustrations, and the use of the words "saddles and lobes," for the purpose of reaching a correct understanding, are of doubtful utility.

This species is distinguished by its general form, transverse section of the volutions and the peculiar sinuosities of the septa.

Found by the late Wm. McAdams in the Coal Measures in Montgomery county, Illinois, and now in the collection of Wm. F. E. Gurley.

GONIATITES KANSASENSIS, n. sp.

Plate V, Fig. 9, lateral view; Fig. 10, surface form of a septum; Fig. 11, ventral view.

Species medium size, subglobose, volutions very slowly enlarging, and lateral and ventral sides regularly rounded. The number of volutions not known. Transverse sections of a volution concavo-convex and the transverse diameter where our shell is broken off is about three times the dorso-ventral. The transverse diameter diminishes towards the apex more than the dorso-ventral and no doubt increases the proportion toward the body chamber of a mature shell. The last volution encloses all the inner ones and leaves a large open umbilicus. The shell is regularly rounded from the open umbilicus, leaving no distinct lateral sides, and the greater transverse diameter near the abrupt walls of the umbilicus. The air chambers are short and complicated. The outer shell is not preserved in our sp·cimen.

Each septum is broadly arched forward from the ambilicus and then curved backward in a waving line to an acute point, posterior to the commencement at the umbilicus, where it makes a sharp retral angle and curves forward in a waiving line slightly anterior to the first forward curve or saddle and then backward in a waiv-

ing line to an acute point, which is slightly anterior to the first
acute point, where it makes another sharp retral angle and again
curves forward in a waving line to a level with the anterior part
of the first saddle and then abruptly curves back a short distance
and then forward and back, so as to make the summit of the saddle bifid, and to form a short, narrow lobe at the periphery in
the middle of the ventral side. There are, therefore, two saddles upon each side of the bifid saddle, as the periphery of
the volution. It will be noticed that the septum above described
is very much like the septa in *G. illinoisensis*, and distinguished
by having shorter saddles, which are less constricted in the middle
part. The sinuosities and shape of the septa in the two species
will be best understood by comparing the illustrations.

This species will be distinguished from *G. illinoisensis* by the
proportionately large umbilicus, shorter dorso-ventral diameter, less
gibbous volutions, which are more abruptly rounded from the
umbilicus, and by the form of the septa. It is probably more
nearly related to that species than to any other which has been
described.

Found by W. J. Parrish in the Upper Coal Measures at Kansas
City, Missouri, and now in the collection of Wm. F. E. Gurley.

GONIATITES GREENCASTLENSIS, n. sp.

*Plate V, Fig. 12, lateral view; Fig, 13, ventral view; Fig. 14,
surface form of a septum.*

Species medium size, globose, volutions expanding laterally quite
rapidly and broadly rounded from umbilicus to umbilicus. The
number of volutions not known. Transverse section of a volution
concavo-convex and the transverse diameter, where our shell is
broken off, is more than four times as much as the dorso-ventral.
The transverse diameter proportionately diminishes toward the
apex and increases toward the body chamber. The last volution
embraces all the inner ones and leaves a large open umbilicus
that is like a hollow cone and formed by the beveling of each
outer volution from the inner volution to the margin of the umbilicus. The shell on the interior of the umbilicus is concentrically lined and furrowed. The shell is depressed convex from
the margin of one umbilicus to the margin of the other, leaving
no lateral sides and the greatest transverse diameter at the margin
of the umbilicus. The surface of the shell is finely cancellated.
The air chambers are rather long.

Each septum is arched backward from the umbilicus to a rather
acute point where it makes a retral angle and curves forward and
back in the form of half an ellipse (but not extending quite as

far anterior as the point of commencement at the umbilicus) and terminates in an acute point at the middle of the ventral side of the volution. There are, therefore, two complete saddles and a half saddle on each margin of each volution and three intervening lobes. The peculiar shape of the septa will be most appreciated by looking at the illustrations.

This species is distinguished by its general form, open concentrically lined umbilicus, flattened volutions, surface ornamentation, and by the form of the septa.

Found in the St. Louis Group at Greencastle, Indiana, and now in the collection of Wm. F. E. Gurley.

GONIATITES SUBCAVUS, n. sp.

Plate V, Fig. 15, lateral view; Fig. 16, end of a volution and ventral view; Fig. 17, surface form of a septum.

Species rather below medium size, subglobose, volutions slowly expanding and broadly rounded from umbilicus to umbilicus. Transverse section of a volution concavo-convex, and the transverse section, where our shell is broken off, is three times as much as the dorso ventral. The transverse diameter proportionally diminishes toward the apex and increases toward the body chamber. The number of volutions not known. The last volution embraces all the inner ones and leaves a large open umbilicus that is like a hollow cone and formed by the leveling of each outer volution, from the inner volution, to the margin of the umbilicus. The shell on the interior of the umbilicus is smooth. The shell is broadly rounded from one umbilicus to the margin of the other leaving no lateral sides and the greatest transverse diameter at the margins of the umbilici. The outer surface of the shell is smooth. The air chambers are very short. Four furrows arise outside of the margin of the umbilicus and curve forward across the ventral side. These furrows do not interfere with the margin of the umbilicus, they are smooth and exist on the outer surface of the shell and on the cast. They do not regularly occur on a volution and belong rather to the outer shell itself than to the body of the volution. It does not appear that they could have had any effect upon the animal.

Each septum arises from the umbilicus and makes a broad curve backward where it turns an obtuse angle and then curves forward nearly as far anterior as the point of commencement and then backward forming more than half an ellipse and again turns an obtuse angle and passes slightly forward to the median line on the ventral side. This forms a bifid lobe and a short narrow saddle in the middle of the ventral side. There are, therefore, five saddles, and four lobes, but the middle saddle and two middle lobes are produced by a bifid lobe.

The septa in this species are very much like the septa in *G. lunatus*, and the open umbilicus is very much like that in *G. greencastlensis*, but the species do not agree in any other respects, and on the whole, have little resemblance to each other. The reader must notice that in looking at figures 5 and 16 he is looking, on the ventral side of the shell toward the apex, and in all other ventral views he is looking toward the anterior end of the shell, and, therefore, the septa in figures 5 and 16 are wrong side up, and the saddles are on the lower side of the septa. Figures 4 and 17 are correct.

Found by the late Wm. McAdams, in the Coal Measures, in Montgomery county, Illinois, and now in the collection of Wm. F. E. Gurley.

GONIATITES JESSIEÆ, n. sp.

Plate V, Fig. 18, lateral view of a small specimen; Fig. 19, end view of a volution and part of a ventral view; Fig. 20, surface view of a septum.

Species medium or above medium size, discoid, sublenticular, volutions rapidly expand, and periphery sharply rounded. We have a specimen more than twice as large as the one that is illustrated, but it shows none of the septa. Transverse section of a volution crescentiform. The sides of the volutions are broadly rounded and the ventral margin more narrowly rounded. The outer volution embraces all the inner ones and closes the umbilicus. The dorso-ventral diameter including the horns of the crescent is about one-half more than the greatest transverse diameter, but the dorso-ventral diameter increases rather more rapidly than the transverse. The external shell is unknown. The air chambers are of medium length.

Each septum curves gently from the umbilicus forward and back, to a point posterior to the place of beginning and near the ventro-lateral margin, where it makes a narrow retral bend and curves forward and backward forming half an ellipse, and then makes a retral bend across the periphery of the ventral side. It is not clear from our specimen whether or not there is a small lobe at the median line. There are, therefore, four saddles and three lobes in each septa as shown by the illustration.

This species is distinguished by its general form, closed umbilicous and surface form of the septa.

Found by R. A. Blair and his accomplished daughter, in whose honor we have proposed the specific name, in the Couteau limestone, near Sedalia, Missouri, and now in the collection of S. A. Miller.

SUBKINGDOM PROTOZOA.

CLASS PORIFERA.

Family RECEPTACULITIDÆ.

RECEPTACULITES DIXONENSIS, n. sp.

Plate V, Fig. 21, basal view; Fig. 22, side view.

Species medium size, general form obovate. Our specimen is more ventricose on one lower lateral side than upon the other. It is a dolomite and the external integument or ectorhim of Billings is not preserved and the internal coating or endorhim is not visible. The part which is presented to us for description is the outer surface of the spicular skeleton.

The base is broadly rounded and has a subcentral, hard, slightly projecting nucleus from which the sigmoidal rows of rhomboidal depressions arise, and curving, at first, gently to the right and to the left, like the engine turnings on a watch, and then curving upward more rapidly, they make more than one revolution around the skeleton before reaching the edge of the summit aperture. All of the rows originate at the margin of the nucleus, and, as they diverge, they increase in diameter, and then contract toward the summit aperture, abruptly stop without the intercalation of any rows. In other words, the surface is covered with the expansion of the rows of rhomboidal depressions that arise at the small solid nucleus, at the base, some of which do not extend to the summit. Each rhomboidal depressions has, within the elevated marginal lines, a transverse furrow with a central pore and one at each end of the furrow. The transverse furrow is crossed by a less conspicious longitudinal furrow. The central pore is larger than the pore at either end of the transverse furrow. The pores and furrows, probably, represent the spicules which formed the skeleton but have been destroyed. The aperture, at the summit, is subcentral but not well preserved in our specimen.

In 1861, Prof. James Hall, in a pamphlet report on the Geological Survey of Wisconsin, page 16, described without illustration a fossil under the name of "*Receptaculites globulars*," as follows:

"Body globuse or subglobuse, with an irregular base of attachment; transverse diameter usually greater than the vertical diameter; summit a little depressed; cells arranged in radiating curved lines, the apertures rhomboidal and transversely elongated; concentric groove and raised ridges between strongly marked. This species is readily distinguished by its small globose form, which is usually not more than three-fourths of an inch in diameter. It is more rare than either of the others (*R. oweni* and *R. iowensis*) though I am informed by Prof. Daniels, that more than twenty specimens were obtained at a single locality in Wisconsin. About twenty years since, I received a specimen of this species from Mr. Thorp, of Mount Morris, Illinois, and have seen others in Galena, and in the collection of Prof. Daniels. Geological formation and locality.—In the Galena limestone of the lead region of Wisconsin, Iowa and Illinois."

The name and definition might have passed into oblivion, because no one could have recognized the species, if Prof. Meek had not revived it, in the Geological Survey of Illinois, vol. 3, p. 301, pl. 2, fig. 2a, b. Prof. Meek described under the name of *Receptaculites globularis*, Hall, a species as follows:

"Body obovate, or subglobose, rounded and slightly umbilicated above, and tapering to a rather broad base of attachment below. Cells arranged in the usual regularly curved lines, with transversely elongated rhomboidal apertures, which become exceedingly narrow and crowded on the sides; transverse ridges between the cells and the intervening grooves well defined, and becoming, like the cells, very closely compacted together on the sides. This is probably the form described by Prof. Hall, under the above name, though it is proportionally longer than the specimens upon which the species was founded, which are said to be usually wider than long. We have others, however, from the same locality agreeing more nearly with his description, and apparently not separable specifically from this. Locality and position.—Scales' Mound, Illinois; from the Galena division of the Lower Silurian series."

We have never seen a specimen that resembles the definition given by Prof. Hall, making due allowance for the fact that he called the summit the base; which was an excusable mistake, until after the study of Billings, on Receptaculites, published in 1865, in Palæozoic Fossils, p. 378. But the species illustrated by Prof. Meek will stand for that of Hall, and we come now to compare it with the species herein described. First, however, we must call attention to the fact that Meek also mistook the base for the summit, and his definition must be corrected in that respect, and his figure 2*a* must be regarded as the base instead of the summit, and figure 2*b* must be reversed end for end. The fact too, that our specimen is much larger than any that either Prof. Hall or Prof. Meek mentioned, is immaterial.

Our specimen is convex at the base, and not umbilicated or concave as *R. globularis* is described. Our specimen does not possess the transversely elongated rhomboidal apertures found in *R. globularis*. And the rows of rhomboidal depressions, in *R. globularis*, as shown in the illustration 2*b*, do not pass half way around the skeleton, while in our species they pass around the skeleton and nearly half around again. The two species, therefore, seem to be widely separated from each other, though they occur in rocks of the same geological age.

It may be proper here to remark, that some European authors widely class American fossils in lists of synonyms with European fossils and with fossils belonging to different geological formations, in America. As an illustration, we find *R. globularis*, which is known only from the Galena Group, in the Lower Silurian, and *R. ohioensis*, and *R. subturbinatus* which are known only from the highest members of the Niagara Group, classed by one of those authors as synonyms for *Ischadites koenigi*. It would seem that some of them have no idea of the order of the geological formations in America, and are equally as obscure in making comparisons of fossils. No species of fossils, animal or vegetable, was ever found common to the Galena and Niagara Groups, and there does not seem to have been any excuse for confounding *R. globularis* with *R. ohioensis* of Meek, or *R. subturbinatus* of Hall, on any palæontological grounds or even upon fanciful resemblance. Neither is there anything in the descriptions or illustrations of *R. ohioensis* by Meek, and *R. subturbinatus* by Hall, that would indicate that they might be synonyms. It will be noticed that Meek, in Ohio Palæontology, Vol. 2, and Hall, in the 11th Report of the Geological Survey of Indiana, continue to call the base, the summit of

—7

Receptaculites. They either overlook the work of Billings, who demonstrated the sponge spicular character of *Receptaculites* or did not choose to follow him in his researches. We think there is no reasonable doubt of the correctness of Billings' observations, on this genus, and adopt his terminology and conclusions.

The type of our species was found in the Galena Group, near Dixon, Illinois, and is now in the collection of Wm. F. E. Gurley.

PLATE I.

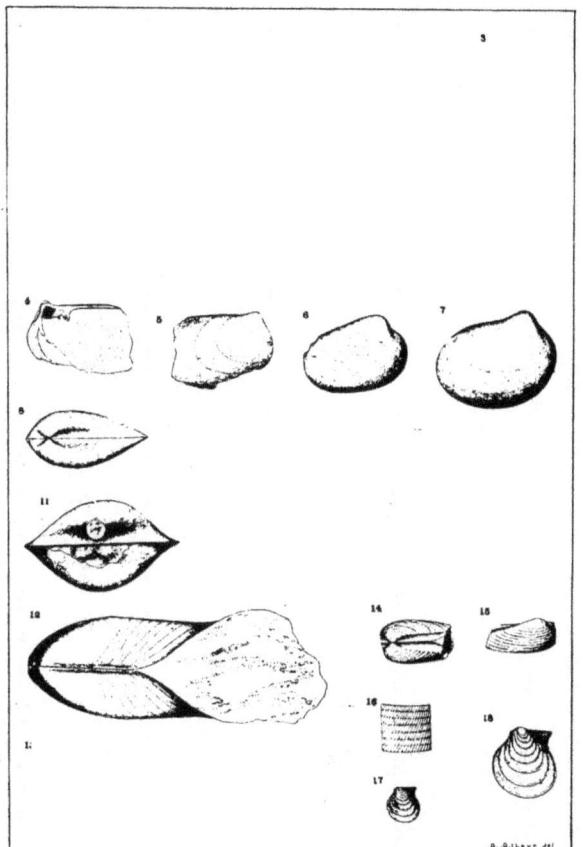

PLATE II.

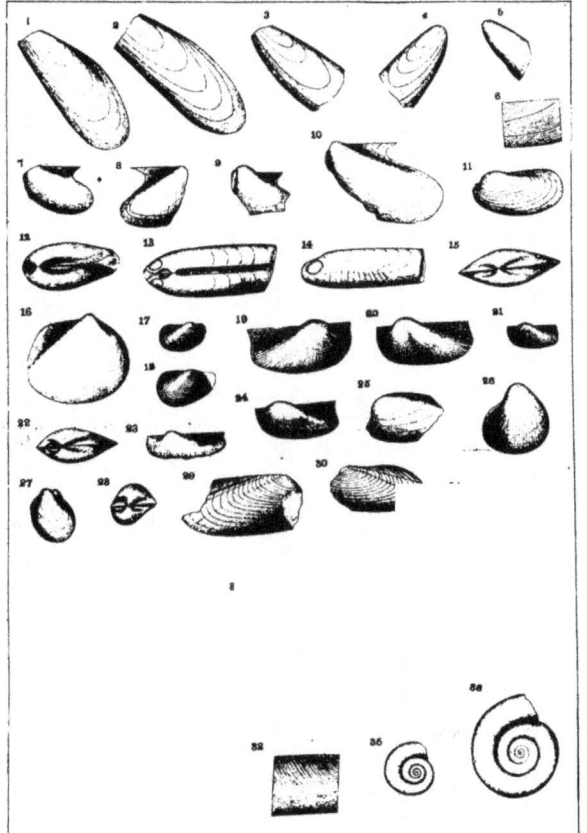

PLATE III.

a.albers. del.

PLATE IV.

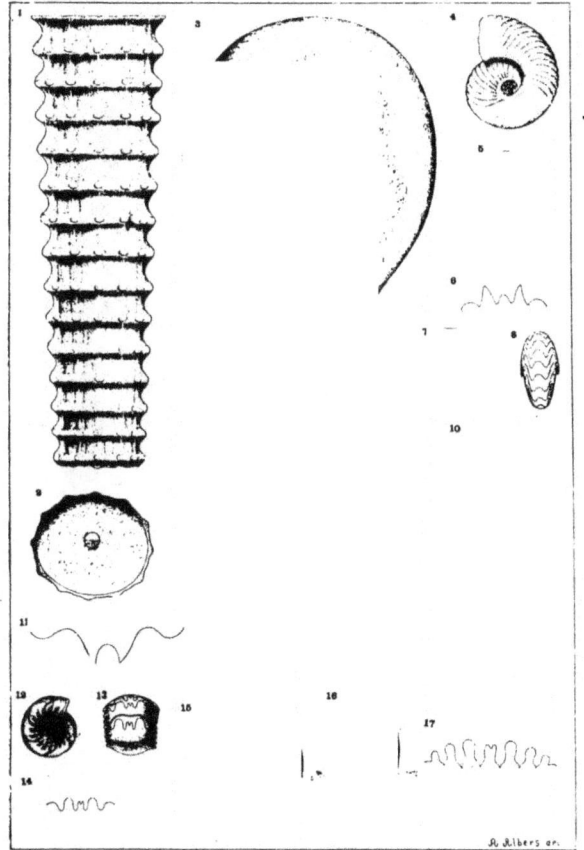

PLATE V.

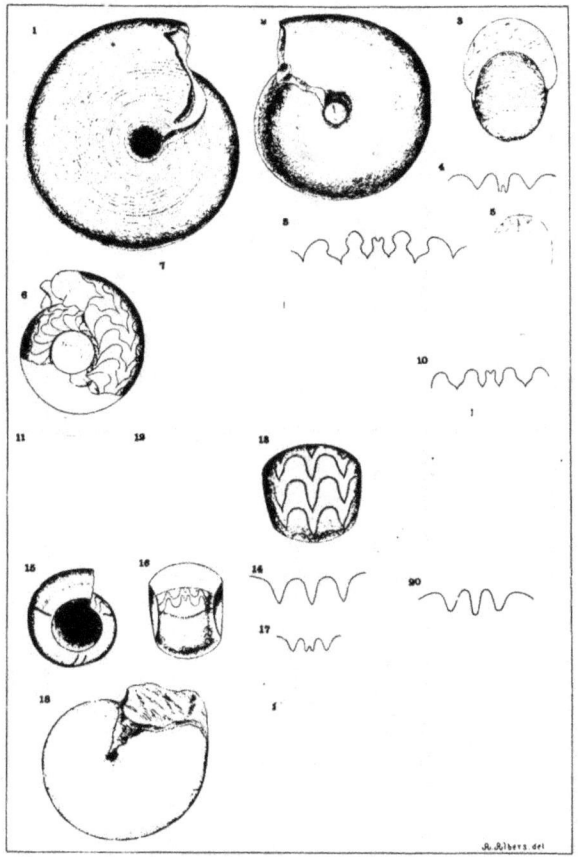

www.ingramcontent.com/pod-product-compliance
Lightning Source LLC
Chambersburg PA
CBHW070933180526
45168CB00003B/1054